To my sons,
Lewis and Arlo.
You mean the most,
my little beans.

- Tim

For Ulysses,
my adorable bean.
Stay curious.

- Kevin

Edited by Lily Coyle
Illustrated and designed by Kevin Cannon
Production editor: Alicia Ester

ISBN 13: 978-1-64343-815-3
Library of Congress Catalog Number: 2021905763
Printed in Canada
First Printing: 2021
25 24 23 22 21   5 4 3 2 1

Beaver's Pond Press
939 Seventh Street West
Saint Paul, MN 55102
(952) 829-8818
www.BeaversPondPress.com

Visit Tim Koerner on Facebook for school visits, speaking engagements, book club discussions, freelance writing projects, and interviews.

WHAT DOES A SCREEN MEAN?

Tim Koerner

illustrated by Kevin Cannon

Beaver's Pond Press
SAINT PAUL, MINNESOTA

Marvin the Bean was born June 13.
His family captured their joy on a screen.

CLICK!

He saw that a screen could show people you love.

He saw it could help Mom pick out the best glove!

On a cold rainy day, his screen offered a game.
He could ask his best friend to play with him, by name!

He could search 'round the world to pick out some adventures.

He could help Grandpa Bean order shiny new dentures!

These days, screens are school,

showing teachers and friends.

Parents use it for work,

but they say "Hope this ends..."

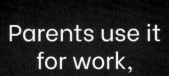

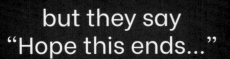

For all those who need it, a screen is a voice.
For Marvin and others, screen time is a choice.

This choice, just like all things, can be good or bad.
Screen time can make you happy, too much can make you sad.

Screens can bring us together or keep us apart,
motivate us or stop us before we can start.

Screens show us great memories or hilarious scenes.
But sometimes, they're where people go to be mean.

What's a screen?
Like a squirt gun: it's a tool we use,
to give joy and do good,
or make sad and abuse.

Screens can open up doors, show new worlds to explore.
But they can keep you trapped.

Today, Marvin the Bean, who was born June 13 is still asking years later, "What does a screen mean?"

He knows screens are just things, no different than others. Sometimes good, sometimes bad, like sisters or brothers.

"What I want from a screen
is fun, laughter, and love.

When my screens bring me down,
I'll just rise above!"

"I'll come back to my screens for occasional fun.
Right now, life's for living. Marvin Bean's got to run!"